Aquatic Plants

Author: Anupam Rajak

Contents

Preface..3

About the Author......................................4

Chapter 1: Importance of Plants..................5

Chapter 2: Aquatic Plants......................6-10

Chapter 3: Aquatic Plants of the World.11-30

References 31-34

Preface

Welcome to Aquatic Plants. This book is helpful to researchers, scholars, students and interested peoples.

Water is the essential part of our life. Without water, we do not live. Plants and animals are made up of water. Without water, you would die. Water is made up of hydrogen and oxygen. Two molecules of hydrogen atom and one molecules of oxygen atom joined together to from water. Human life need clean water. There are two kinds of water present on Earth- salt water and fresh water. The various functions of water in plant life is maintaining cell growth, transport nutrients, photosynthesis, transpiration, and germination of seeds.

Aquatic means existing or happening of water. Aquatic plants are grows in water. They are found in seawater or freshwater. They grow in ocean, river, lake, pond, shorelines, and many more places. Aquatic plants are also called as hydrophytes or macrophytes. They are distributed all over the World.

The first chapter briefly describes importance of plants. The second chapter gives definition, character, and types of aquatic plants. The third chapter covers various aquatic plants includes Pistia stratiotes, Eichornia crassipes, Lemnoideae, Isoetes, Salvinia molesta, Nelumbo nucifera, Pygyma Wter-lily, Victoria amazonica, Nymphaea alba, Nuphar japonica, Euryale ferox, Azzola filiculoides, Nymphaea thermarum, Ceratophyllum submersum, Lemna trisulca, Wolffia arrhiza, Hydrilla verticillata, Valisneria, Gymnocoronis spilanthoides, and Potamogeton. The third chapter also describes importance of aquatic plants.

If any error found this book please mail to anupamrajak1234@gmail.com

About the Author

Anupam Rajak received his B.Sc in Botany from the Raghunathpur College, Sidho-Kanho-Birsha University. He has published several articles in international journal. His email address is anupamrajak1234@gmail.com

Chapter 1

Importance of Plants

Plants are photosynthetic organisms. Plants are multicellular. Plants are autotrophic to make their own food through photosynthesis. Plants have vascular tissue. Plants reproduce both sexually or asexually.

Importance:

Without plants, we donot live on the Earth. Plants are very important for our life. Plants absorb carbon dioxide and release oxygen into the environment. Plants provide us to food, medicine, oxygen, wood, fibres, pesticides, oils, rubber, and more essential things.

Figure 1.1 Aquatic Plants (Photo Credit: Pixabay)

Chapter 2

Aquatic Plants

Water is the essential part of our life. Without water, we do not live. Plants and animals are made up of water. Without water, you would die. Water is made up of hydrogen and oxygen. Two molecules of hydrogen atom and one molecules of oxygen atom joined together to from water. Human life need clean water. There are two kinds of water present on Earth- salt water and fresh water. The various functions of water in plant life is maintaining cell growth, transport nutrients, photosynthesis, transpiration, and germination of seeds.

Aquatic means existing or happening of water. Aquatic plants are grows in water. They are found in seawater or freshwater. They grow in ocean, river, lake, pond, shorelines, and many more places. Aquatic plants are also called as hydrophytes or macrophytes. They are distributed all over the World.

Figure 2.1 Aquatic Plants (Photo Credit: Pixabay)

Characteristics of Aquatic Plants:

i. Aquatic plants have thin cuticle. Submerged plants lack cuticle.

ii. Leaves bears a number of stomata. Submerged plants lack stomata.

iii. Some aquatic plants float on water while other live in water.

iv. They have a air filled cavities in their leaves and stomata.

v. Leaves are flat.

vi. Roots are feathery and small.

Types of Aquatic Plants:

Aquatic plants are divided into three categories-

i. Free Floating Plants.

ii. Submerged Plants.

iii. Emergent Plants.

i. **Free Floating Plants:** They are float on the water surface. Some of them are rootless. Their leaves and flower are freely moves and float on the water surface. Examples- water lily, water lettuce, and duckweed.

Figure 2.2 Duckweed (Photo Credit: Pixabay)

ii. Submerged plants: They are grow underwater. Examples:- hornwort, nymph.

Figure 2.3 Nymphaeae alba (Photo Credit: Pixabay)

iii. Emergent plants: Emergent plants have rooted. They grow along the shore where water is low. They are also called as shorelines and wetland plant. Emergent plant live in water-lodged soil. Examples- Cattails, Tape grass.

Figure 2.4 Cattails (Photo Credit: Pixabay)

Chapter 3

Aquatic Plants of the World

Pistia stratiotes:

Pistia stratiotes is an aquatic plant, which belongs to the family Araceae and order Alismatales. Pistia stratiotes float on the surface of the water. Pistia stratiotes is commonly called as water cabbage, water lettuce, or shellflower. They are found in south and Central America, Africa, Europe, North America, Oceania, and Asia. P.stratiotes is a perennial monocotyledonous aquatic plant. Leaves are pale green and without petioles. Flowers are unisexual.

They are used to animal food. They are ornamentally grow in botanical garden, and zoo. They are used to treat various disease.

Figure 3.1 Pistia stratiotes (Photo Credit: Pixabay)

Eichornia crassipes:

Eichornia crassipes is a free floating aquatic plant, which belongs to the family Pontederiaceae and order Commelinales. Eichornia crassipes is commonly known as water hyacinth. They are found in South America, Africa, Austrilia, India, and many more countries. Leaves are oval to elliptical. Leaves consists of a petiole and blade.

Figure 3.2 Eichornia crassipes (Photo Credit: Pixabay)

Lemnoideae:

Lemnoideae is an aquatic plant, which belongs to the family Araceae and order Alismatales. Lemnoideae is commonly known as duckweed, water lentils, or water lenses. Duckweed is floats on the surface of the water. They are used as food.

Figure 3.3 Lemnoideaea –Duckweed pond green (Photo Credit: Pixabay)

Isoetes:

Isoetes is an aquatic plant, which belongs to the family Isoetaceae and order Isoetales. Isoetes is commonly known as 'quillworts'.

Figure 3.4 Isoetes (Photo Credit: Flickr)

Salvinia molesta:

Salvinia molesta is an aquatic fern, which belongs to the family Salviniaceae and order Salvinales. Salvinia molesta is commonly known as giant salvinia or kariba weed. They grow on the surface of a body of water. It is found in north central, and southwest Florida.

Figure 3.5 Salvinia molesta (Photo Credit: Flickr)

Nelumbo nucifera:

Nelumbo nucifera is an aquatic plant, which belongs to the family Nelumbonaceae and order Proteales. Nelumbo nucifera is commonly known as indian lotus, sacred lotus, Egyptian bean, or lotus. They grow in pond or river water.

Figure 3.6 Nelumbo nucifera (Photo Credit: Pixabay)

Pygmy water-lily:

Pygmy water-lily is belongs to the family Nymphaceae and order Nymphaeales.

Figure 3.7 Pygmy water-lily (Photo Credit: Pixabay)

Victoria amazonica:

Victoria amazonica is the National flower of Guyana. Victoria amazonica is the largest flowering plants of Nymphaeaceae family. Victoria amazonica has large leaves and float on water. The flowers are white or pink appearance. Flowers are pollinated by scarab beetles.

Uses:

i. They are used as ornamental plant.

ii. They are used as food.

Figure 3.8 Victora amazonica (Photo Credit: Pixabay)

Nymphaea alba:

Nymphaea alba is an aquatic floweing plant, which belongs to the family Nymphaceae and order Nymphaeales. Nymphaea alba is also known as European white water lily, or white nenuphar. They are found in Europe, North Africa, Middle East, and tropical Asia. It grows in water. The flowers are white.

Figure 3.9 Nymphaea alba (Photo Credit: Pixabay)

Nuphar japonica:

Nuphar japonica is an aquatic plant, which belongs to the family Nymphaceae and order Nymphaeales. Nuphar japonica is also known as East Asian yellow water lily. It is found in Japan, and Korean peninsula.

Figure 3.10 Nuphar japonica (Photo Credit: Flickr)

Euryale ferox:

Euryale ferox is a flowering plant, which belongs to the family Nymphaceae and order Nymphaeales. Euryale ferox is commonly known as brickly waterlily, foxnut, or gorgon nut. They are found in China, Japan, Korean, India, Bangladesh, and Myanmar. They are used as food.

Azzola filiculoides:

Azzola filiculoides is a water fern, which belongs to the family Salviniaceae and order Salvinales. It is found in America, Asia, and Austrilia.

Figure 3.11 Azzola filiculoides (Photo Credit: Wikimedia commons/ I, Daniel J. Layton / CC BY-SA (http://creativecommons.org/licenses/by-sa/3.0/))

Nymphaea thermarum:

Nymphaea thermarum is belongs to the family Nymphaceae and order Nymphaeales.It is the smallest waterlily in the World. It is found in ,ashyuza Rwanda (Africa). The flowers is white.

Figure 3.12 Nymphaea thermarum (Photo Credit: Wikimedia Commons/ C T Johansson / CC BY-SA (https://creativecommons.org/licenses/by-sa/3.0))

Ceratophyllum submersum:

Ceratophyllum submersum is a submerged, free floating aquatic plants, which belongs to the family Ceratophyllaceae and order Ceratophyllales. It is commonly known as the soft hornwort or tropical hornwort. It is found in Europe, Central Asia, northern Africa, Turkey, Oman, Florida, and Dominion Republic.

Figure 3.13 Ceratophyllum submersum (Photo Credit: Wikimedia Commons/ Totodilefan / Public domain)

Lemna trisulca:

Lemna trisulca is an aquatic plant, which belongs to the family Araceae and order Alismatales. It is commonly known as star duckweed, or ivy-leaved duckweed. They are found in Great Britain, Ireland, Asia, India, Indonesia, Brunei, Japan, and Europe. It floats on the surface of water.

Figure 3.14 Lemna trisulca (Photo Credit: Wikimedia Commons/ Lamiot / CC BY-SA (https://creativecommons.org/licenses/by-sa/4.0))

Wolffia arrhiza:

Wolffia arrhiza is an aquatic plant, which belongs to the family Araceae and order Alismatales. Wolffia arrhiza is commonly known as water meal and rootless duckweed. It is evergreen perennial plant. This plant lacks root. It floats on water bodies like ponds.

Figure 3.15 Wolffia arrhiza (Photo Credit: Wikimedia Commons)

Hydrilla verticillata:

Hydrilla verticillata is an aquatic plant, which belongs to the family Hydrocharitaceae and order Alismatales. Hydrilla verticillata is a submersed and herbaceous perwnnial plant. They grow in water bodies.

Figure 3.16 Hydrilla verticillata (Photo Credit: Dreamstime)

Vallisneria:

Vallisneria is an aquatic plant, which belongs to the family hydrocharitaceae and order Alismatales. Valisneria is commonly called eelgrass, or tapegrass. Vallisneria is distributed all over the World. Generally, they are found in Asia, Africa, Europe, and North America.

Figure 3.17 Vallisneria (Photo Credit: Wikimedia Commons/ Damitr / CC BY-SA (https://creativecommons.org/licenses/by-sa/4.0))

Gymnocoronis spilanthoides:

Gymnocoronis splanthoides is an aquatic plant, which belongs to the family Asteraceae and order Asterales. Gymnocoronis is come from two Greek words 'Gymnos' mean 'naked' and 'corona' means 'crown'. They are found in Asia, Africa, Europe, Oceania, and South America. They are used as food.

Figure 3.18 Gymnocoronis spilanthoides (Photo Credit: Wikimedia Commons/ Krzysztof Ziarnek, Kenraiz / CC BY-SA (https://creativecommons.org/licenses/by-sa/4.0))

Potamogeton:

Potamogeton is an aquatic plant, which belongs to the family Potamogetonaceae and order Alismatales. Potamogeton is come from two Greek words 'Potamos' means 'river' and 'geiton' means 'neighbour'. Leaves submersed and floating.

Figure 3.19 Potamogeton (Photo Credit: Flickr)

Importance of aquatic plants:

Aquatic plants have many important functions. Some are described below-

i. Aquatic plants remove waste, and decaying matter.

ii. Aquatic plants absorb nitrates through leaves.

iii. They support the life cycle of many fish and shellfish.

iv. They produce leaves and stems that is valuable for food sources.

v. They produce oxygen through photosynthesis.

Figure 3.20 Aquatic Plants leaves (Photo Credit: Pixabay)

References:

1. https://en.wikipedia.org/wiki/Aquatic_plant

2. https://www.toppr.com/guides/biology/plants-and-mushrooms/aquatic-plants/

3. https://www.proflowers.com/blog/aquatic-plants-and-flowers

4. http://www.aquarius-systems.com/Pages/86/common_aquatic_plants.aspx

5. https://www.1800flowers.com/blog/flower-facts/all-about-aquatic-plants/

6. https://aquaplant.tamu.edu/plant-identification/category-emergent-plants/

7. https://www.sorkoservices.com/2018/11/19/what-are-emergent-plants/#:~:text=Emergent%20aquatic%20plants%20grow%20in,leaves%20rising%20above%20the%20water.

8. https://www.lakedoctors.com/emergent-plants/

9. https://gobotany.nativeplanttrust.org/simple/aquatic-plants/

10. https://www.biologydiscussion.com/plants/xerophytes/hydrophytes-meaning-and-characteristics-plants-botany/75467

11. https://en.wikipedia.org/wiki/Pistia

12. https://www.cabi.org/isc/datasheet/41496

13. Tripathi, P et al. "Pistia stratiotes (Jalkumbhi)." *Pharmacognosy reviews* vol. 4,8 (2010): 153-60. doi:10.4103/0973-7847.70909

14. https://en.wikipedia.org/wiki/Eichhornia_crassipes

15. https://www.cabi.org/isc/datasheet/20544

16. https://wiki.bugwood.org/Eichhornia_crassipes

17. https://en.wikipedia.org/wiki/Lemnoideae

18. https://pondmegastore.com/pages/what-is-duckweed-lemnoideae

19. https://en.wikipedia.org/wiki/Salvinia_molesta

20. https://plants.ifas.ufl.edu/plant-directory/salvinia-molesta/

21. https://en.wikipedia.org/wiki/Nelumbo_nucifera

22. http://fieldguide.mt.gov/speciesDetail.aspx?elcode=PDNYM050J0

23. https://en.wikipedia.org/wiki/Victoria_amazonica

24. https://www.worldatlas.com/articles/amazing-facts-about-the-victoria-amazonica.html

25. https://anuragdhyani.com/victoria-amazonica-the-largest-water-lily-in-the-world/

26. https://en.wikipedia.org/wiki/Nymphaea_alba

27. https://indiabiodiversity.org/species/show/230482

28. https://en.wikipedia.org/wiki/Nuphar_japonica

29. https://www.lilieswatergardens.co.uk/nuphar-japonica-barerooted-p-3336.html

30. https://en.wikipedia.org/wiki/Euryale_ferox

31. http://tropical.theferns.info/viewtropical.php?id=Euryale+ferox

32. https://en.wikipedia.org/wiki/Azolla_filiculoides

33. http://tropical.theferns.info/viewtropical.php?id=Azolla+filiculoides

34. https://www.cabi.org/isc/datasheet/8119

35. https://wiki.bugwood.org/Hydrilla_verticillata

36. https://en.wikipedia.org/wiki/Vallisneria

37. https://www.cabi.org/isc/datasheet/26246

38. https://en.wikipedia.org/wiki/Water

39. *Water Is Life - Texas Aquatic Science - Rudolph Rosen.* 6 June 2019, texasaquaticscience.org/water-texas-aquatic-science/.

40. *Functions of Water in Plants,* www.personal.psu.edu/faculty/a/s/asm4/turfgrass/education/turgeon/lessons/lesson05/corefiles/links/effectsmoist/1.html.

41. http://agritech.tnau.ac.in/agriculture/agri_irrigationmgt_roleofwater.html

42. Info, Author Aquarium. "Top 6 Benefits of Aquatic Plants in the Aquarium." *Aquarium Info*, 26 Jan. 2016, blog.aquariuminfo.org/top-6-benefits-aquatic-plants-aquarium/.

43. US Department of Commerce, National Oceanic and Atmospheric Administration. "Why Are Aquatic Plants so Important?" *NOAA's National Ocean Service*, 1 June 2013, oceanservice.noaa.gov/facts/underwaterplants.html.

44. https://en.wikipedia.org/wiki/Potamogeton

45. http://www.efloras.org/florataxon.aspx?flora_id=1&taxon_id=126627

46. https://en.wikipedia.org/wiki/Isoetes

47. http://powo.science.kew.org/taxon/urn:lsid:ipni.org:names:936393-1

48. https://discoverandshare.org/2019/05/30/worlds-smallest-water-lily-nymphaea-thermarum/

49. http://www.watergardenersinternational.org/journal/4-4/carlos/thermarum_gallery1.html

50. https://en.m.wikipedia.org/wiki/Ceratophyllum_submersum

51. https://en.m.wikipedia.org/wiki/Lemna_trisulca

52. https://www.illinoiswildflowers.info/wetland/plants/star_duckwd.html

53. https://en.m.wikipedia.org/wiki/Wolffia_arrhiza

54. https://www.dntaqua.com/aquarium-plants/wolffia-arrhiza/

www.ingramcontent.com/pod-product-compliance
Lightning Source LLC
Chambersburg PA
CBHW040340220526
45473CB00009B/2748